THE NORTHWEST ARCHITECTURE

of *Pietro Belluschi*

THE NORTHWEST ARCHITECTURE

of *Pietro Belluschi*

Edited by JO STUBBLEBINE

AN ARCHITECTURAL RECORD BOOK

F. W. Dodge Corporation, New York

Library of Congress Catalog Card No. 53–7216

CONTENTS

FOREWORD

This book was undertaken at the time when Pietro Belluschi left his adopted section of the Pacific Northwest to become Dean of the Massachusetts Institute of Technology Department of Architecture and Planning. It seemed the end of a chapter in his work, with many more sure to follow, as he continues to use the tools he has perfected these past two decades. His influence will always be felt in the progress of an indigenous American architecture in the Northwest vernacular. He has left many buildings and homes as expressions of his ideals and development, as well as many younger architects who have caught his spirit of environmental design.

The purpose of this collection of photographs is to show his versatility, his understanding of the character of the land, its people, its materials, and his application of progressive ideas.

The text of the book is made up of Belluschi's own speeches and published statements describing his philosophy and beliefs. These were made over a long period of time, to a variety of audiences, but hold present-day validity.

If more complete studies of his works are desired, a bibliography of publications, in which technical details are available, is to be found at the end of the book.

My appreciation is hereby expressed to those who have assisted me in the preparation of this material. Mr. Belluschi has given me full cooperation and assistance; his office staff has been encouraging and helpful. In particular, I wish to thank Kenneth E. Richardson, architect and photographer, whose guidance in the selection of photographs was indispensable.

Jo Stubblebine

Ray Atkeson

ABOUT BELLUSCHI HIMSELF

By JO STUBBLEBINE

ABOUT BELLUSCHI HIMSELF

Pietro Belluschi was born in Ancona, Italy, a small town on the Adriatic Sea near Venice. When he was six his family moved to Rome, where he spent most of his youth. Of his early years he has this to say, "Rebellion against my middle-class environment gave me a stronger motivation for artistic expression than the pervading beauty of Rome. My desire was to differ, not to emulate; so I had no hero but my own self.

"Drawing was perhaps the only thing I ever did in school with any amount of pleasure. My interest in architecture began in high school, but the inner urge to create came when the ways of rebellion were shown to me. The early iconoclasts struck a deep chord within me; the shackles created by history were to be broken; it was a task worth going after."

1917

At 17 Belluschi volunteered in the Italian Army, then at war, and served three years in the mountain artillery. After his return to Rome he attended the University, and received his Doctor of Architectural Engineering degree from the School of Application for Engineers in 1922. An exchange graduate scholarship to Cornell University, granted by the Italy-America society of New York, brought him to America in the fall of 1923.

1924

"Coming to America changed my life to a greater extent than I thought possible. From a dreamy, lazy boy I became almost overnight an aggressive and determined man—determined to

1

HARRY WENTZ

WENTZ COTTAGE

HOUSE SHOWING JAPANESE INFLUENCE (See page 5)   *Boychuch*

succeed at all costs as a student, as a person, and as an architect."

Upon receiving his civil engineering degree at Cornell, he decided to use what money he had left to see the West and continue to learn the language. On the advice of the Italian Ambassador he procured a job in the mines of Idaho as a helper electrician. After a time, he continued on to Portland, Oregon, seeking an architectural job. He joined the old and respected firm of A. E. Doyle in April, 1925.

1 9 2 7

In addition to his hard work and natural talent, his career was favored by circumstances. Mr. Doyle's death and the coinciding departure of other members of the firm for various reasons, left the way open for Belluschi to advance to head designer in two years following his arrival. For fifteen years he held this position under the Doyle name, until he acquired the firm and changed to his own name in 1943.

In his early days, when his interest in design was flowering, he studied and worked with great intensity toward a goal of expressive simplicity. He was greatly influenced by the philosophy of Harry Wentz, well-known artist and teacher at the Portland Museum Art School. As Mr. Wentz says, "From his Italian background he brought a deep interest in *all* the arts. He had a desire to decorate with significant pattern and at times applied it, if not in forms, in color."

He kept his love of marble, which became almost a trademark in his public buildings. His extensive use of loggias and open courts is easily justified in Oregon's mild climate, but is reminiscent of Italian planning. His desire to give emotional appeal is evident in the many times sculpture and embossed copper have been integrated into his designs. He feels architecture is the greatest of all the arts, because it has a social purpose and is a base for all other art.

Belluschi, John Yeon, and other young architects trying to clarify their thinking spent much time with Mr. Wentz during the depression of the '30s, on sketch trips and in earnest discussion of design philosophy. They shared the inspiration of the countryside, the forests, and the early buildings of this north-

Ray Atkeson

K. E. Richardson

Ray Atkeson

west country. Although Belluschi preferred his native sea to the cold Pacific, he had a keen enthusiasm for the lush greenery, the towering trees, and the stately old barns of his adopted section.

Impressed by the Neahkahnie beach cottage of Mr. Wentz, which was built in 1916, he went on to make a reputation for his "Oregon" house, or "beautiful barns," as many called them. These were noted for the use of natural, native woods and a particular feeling for their environment. The Wentz studio cottage is considered by most of the architects of this region as being the prototype or beginning of Northwestern contemporary domestic architecture. It was planned by Wentz and executed by A. E. Doyle. Mr. Humke, a local craftsman, built it alone, with occasional help from Mr. Wentz. Depending on his eye and imagination more than his square and level, he lent it a handmade look; a variety and charm. Mr. Belluschi says of the cottage, "It has function, appropriateness, harmony, materials, setting, orientation; it is modern, emotional, beautiful." As more and more of his own houses reflected these qualities, he gained national and international fame as one of America's outstanding regionalists.

The Japanese influence certainly cannot be denied. The low horizontal lines, the sheltering roofs, the restraint of detail, the significance of the garden relationship to the house, the rhythms of all components, bespeaks the taste and aesthetic quality of Japanese architecture.

What Belluschi saw in Europe when he made a visit home in 1934 was more than the flavor of French modern and the new mode of Italian architecture; it was the appropriateness of the old medieval villages and churches, born of their time and culture. This awareness accelerated his desire to design for the present time and way of life, to help in the development of a true, indigenous architecture.

His generation was rightly timed to pick up and apply what had been stirring previously. The work of Frank Lloyd Wright was encouraging and inspiring Belluschi and others like him. Although many believe that contemporary architecture thrives in areas where there is indifference rather than enlightenment,

K. E. Richardson

STEARN MEMORIAL FOUNTAIN

Erven Jourdan

(Above) ST. THOMAS MORE CHAPEL   (Below)   COTTAGE GROVE PRESBYTERIAN CHURCH

Julius Shulman

6

Belluschi was skillful in educating clients to let him materialize progressive ideas. His warm Italian personality and sincere approach to a problem has served to convince and convert many who were hesitant.

"In my advance, I was to be helped more by my intuitive sense and critical sense of fitness than by my creative abilities, which were limited. My later professional success was due not a little to luck, much to a practical business approach coming from my early middle-class influence, and to an ability to deal with people."

He was extravagant with his design time and effort, he was adept at sensing the needs—and significantly—the feelings of his clients.

Recognition in the field of public buildings began with the rosy brick and travertine of the Portland Art Museum, designed in 1931. The simplicity without starkness, and the easy proportions won a place for it among the 100 best buildings in the United States in the period from 1920 to 1940, as chosen by the Committee of Education of The American Institute of Architects. Finley's Mortuary was also included, and from these, through a series of smaller commercial buildings, downtown remodeling, eye-catching shop fronts, and the massive efficiency of the Oregonian Building, he progressed to a masterpiece of engineering and design—the Equitable Building.

Several outstanding churches have come from his drawing board. The first, in 1938, was St. Thomas More chapel, with its spire over the chancel and its straightforward use of wood. After a long lapse, in which Belluschi's philosophy underwent a struggle against the materialism and spiritual poverty of man, have come three churches, full of spiritual feeling and with an elegance of simple interplay of brick, glass and wood. The First Presbyterian Church at Cottage Grove, Oregon, with its unique roof line, is described by Belluschi in his own words, "Because the aims and reasons expressed by the architects were understood and wholeheartedly shared by the pastor and the congregation, the design of their church has been a memorable experience. Here was a fine lot, bordered by beautiful trees, located in the midst of the residential district, and presenting wonderful opportunities to create an intimate and inviting atmosphere. With that purpose in mind, a landscaped forecourt was introduced to provide a zone of quiet, a sort of transition area designed to dispose the churchgoer inwardly, and to create a feeling of space and expectation, both of which are such important and subtle elements in architecture.

"The materials used are humble ones, and the details very simple, chosen more to convey the idea of purpose than that of richness, and to prove that architecture is an intrinsic art and not an arbitrary dress to be applied at the designer's changing whim."

Herman Brookman, while president of the Oregon Chapter of The American Institute of Architects, made the following remarks on the occasion of Mr. Belluschi receiving an honorary LLD from Reed College on January 17, 1951:

"There has never been any doubt whatever in the minds of us architects that by his brilliant talent, hard work, uncompromising effort, and with that remarkable good taste and common sense, that Pietro would reach the pinnacle of our profession.

"Belluschi's projects fortunately are solved with great simplicity. I believe that his approach to all problems that are constantly before us in our profession, is to choose the simplest solution. . . . . His sense of scale, handling of material and color is a stimulation to the imagination.

"The many offices he has held in our Chapter and at present holding in the Institute in Washington, are but recognition of his ability as an outstanding designer and leader of our profession.

"Many honors have come to Pietro during the last few years; as President of our local Chapter; the award of Fellowship by the Institute [1948—cited for his contributions to design, to public service, and to the A.I.A.]; the appointment by the President of the United States as member of the National Commission on Fine Arts [1950]; and now chosen as Dean of the Massachusetts Institute of Technology's School of Architecture and Planning.

"We know that Pietro with his warm and sympathetic nature and gift of interpreting ideas will instill in the hearts and minds of his students a hunger and love for genuine beauty."

8

PART II

IN BELLUSCHI'S WORDS

Ray Atkeson

ARCHITECTURE AND SOCIETY

Reed College, 1951, P. Belluschi.

I am glad to have lived through such a momentous period of change. My early life was influenced and not a little stifled by the glory that was Rome; as a student in architecture I could hear the wide discontent of the intellectuals and their desire for renovation. Led by a few men of genius, the younger generation

10

was spoiling to shake architecture out of its lethargy, slay the "Beaux Arts" dragon, and clear the ground for the new era.

Of the ideals on which contemporary architecture is based, first and foremost is the right to free our thinking from the dogmas of the past. It has now become clear that the various historical forms developed by past generations cannot serve us well. Freedom has its dangers; without discipline it leads to anarchy—but just as in politics, freedom is the healthiest climate for progress.

Complementing this ideal of freedom is the right to interpret our own world in forms suitable to the demands and purposes of the times. We believe that architecture, in order to be significant, must absorb and give meaning to modern methods of construction, and to newly developed materials, as well as reflect the physical environment of a region and particularly the traits of its people. In this respect, the west coast—with the pioneering heritage of its people, with definite natural characteristics of its own, and with less binding ties to the past—has been able to advance more visibly towards the realization of valid contemporary forms.

Finally, we believe in establishing the right to think and speak in behalf of our own society, if we can ever hope to be of help in bringing some degree of order out of the confused and ugly environment which is the modern city. In this important task, conservative architects have looked in vain to the past for ready-made solution, using as example and guide the European city, with its stately palaces, fountains, monuments and plazas. Unfortunately, the social order which produced such appealing forms no longer exists.

Our own society is conditioned by the machine and dominated by the desires of the common man. The common man no longer wants to live in slums; he does not ask for stately palaces but for clean houses and children's playgrounds. He wants comfort at the factory and recreation after work; he wants good schools and good transportation; he demands that the problems created by traffic, smoke, parking, and shopping be solved to his convenience—in brief, he wants an efficient city, and in this he is right. Surface embellishments may then come later at a time when our aesthetic creativeness will have reached its full maturity.

The ideals of the modern man may then be very briefly summarized as follows: He must come to terms with his environment; only thus can an architect hope to become again creative, not in the anemic method of the academy, or as a fashionable hireling of the wealthy, but as a lively interpreter of the new social order and as a prophet of his age.

To what extent have we succeeded up to now? We readily admit that our accomplishments are very modest, and our successes mostly on the negative side. What little we have to show for our efforts has not been easily achieved, not so much because of the doubters among clients and public, but mostly because of our own conflicts and limitations. We had to find our way among the great many technical advances, and distinguish the basic from the superficial; we had to develop the inner discipline which alone could prevent us from being seduced by the many transitory forms offered for daily consumption. It is also apparent that we have succeeded in designing good factories but have failed to create beautiful monuments. Today we are more honest, more practical, and quite functional, but it has been at the expense of grace and gentility. We have taken away many of the established forms, so cherished by our ancestors, and have replaced them with stark utilitarian ones, which give little nourishment to the senses. We have taken away from the man in the street all the stereotyped little ornaments, cornices, cartouches and green fake shutters, but we have not been capable of giving him back the equivalent in emotional value. The fact is, that after three decades of rather cold functionalism, we have come to the realization that emotion is a great force in our everyday world; it pervades our actions, our political motives, our very happiness—yet emotions have not been given the guidance they deserve, although they are the very soil in which both architects and public may grow to creativeness and understanding.

By looking at our cities it is quite obvious that we have not been the interpreters and the prophets we had wished to be; we are still shy on wisdom but I believe our thinking has acquired a greater clarity of purpose and discovered new aspects of beauty, yet to be translated. We have also found that beauty is forever changing and eluding possession, perhaps because of the power of the human mind to perceive and to create, and that power has no end. We have rediscovered on our own terms that architecture is the art and science of organizing space and relating it to man for his pleasure and comfort, and that an architectural work really lives and shines only when it is part of a larger organization.

It may be said that the sum total of our vision spells "Utopia," but I believe that the complex events of our modern life which eventually will force us to make fundamental decisions are accelerating in their tempo. Wars, obsolescence, traffic, air travel, mass education, and so on, will inevitably bring us new demands for change and from them new forms. If we are prepared, and if our vision is clear, we can make each move—however small—an orderly and logical step toward the total plan.

When I compare what was produced in the architectural schools years ago, when the Beaux Art held power, and when all good architects came from Paris, with the present work done by today's students, I feel greatly encouraged.

I believe that the next generation will make us really proud; from the lesson we have learned I hope they will acquire a new discipline of the mind to take the place of the discipline of the "styles," and that they will have enough feeling and integrity of purpose to make their work of lasting significance.

And now that most of the battles against dogmas have been won, I hope they may also gain a certain amount of tolerance for all the human symbols and forms of the past, because people need them and live by them to a greater extent than is realized, because they furnish a feeling of continuity which gives them faith in their evolution. This fact the architects must understand if they want to be the leaders.

In these dark times we have a greater need of faith in the future than ever, but I persist in the optimistic view that in all events the foundations of a new renaissance are being laid now. It will not be for us to see it, and we must only reckon in terms of generations for its flowering, but I believe a better environment for a happier mankind is in the making. It is a task to excite the imagination, and it is now in the hands of our young people.

Ray Atkes

THE ROLE OF THE MODERN MUSEUM IN OUR NEW CIVILIZATION

Pietro Belluschi, Lecture, Portland Art Museum, 1935.

It is easy to understand why museums are a serious business in the European countries when you consider the wealth of their accumulated treasures and the direct and indirect importance of these treasures in the balance of their economy.

A modern and alive museum, at least in this country, does not merely collect old paintings for the sake of showing how wealthy is a city or how much pseudo-culture exists in its population, but it should attempt above all to meet squarely present conditions of living, realizing that men need production of materials on a great scale, but also that material things to serve us completely must serve our thoughts and feelings as well as sensations and so come into the field of art, a field admitting no hard and fast division between the fine and the practical arts.

The museum's most important mission is furnishing the means of restoring aesthetics to their normal relation to the other human activities—the mission of reviving the power of expressing and integrating the arts with the totality of our life experience, of making things not only usable but expressive of life, richness and unity.

The museum must serve its high educative purpose, must reinstate the artist to his main role of prophet, make him the articulate soul of mankind and not just a vapid maker of ornamental "artiness."

The museum is in a deep sense the mirror of the civilization which built it, yet it must lead it in helping to find its realization and its understanding. It must be animated by perpetual motion and feel the creative pulse of the city and the nation and the world. To create a culture not forgetful of the past, but more concerned with the present; to create a culture more deeply concerned with life, is the purpose of museums.

Ray Atkeson

CHURCH ARCHITECTURE

Pietro Belluschi, Liturgical Arts, November, 1950.

In approaching the problem of designing religious buildings, the contemporary architect is confronted by the difficult problem of creating form appropriate to a modern society without destroying the many symbols which have given formal validity to the idea of a *church* in the past. These symbols, crystallized through the centuries, have become identified in the minds of many with religious belief itself, and they give much strength to religious institutions, particularly the Catholic Church. The extent to which we can preserve them and still speak the language of our own time is the real problem confronting us. The modern architect has found that his integrity will prevent him from building with the tools of the past, or to use deception in forcing old architectural forms onto modern materials; yet he has found that he must respect and preserve that feeling of emotional continuity which is the very essence of religion.

[It has been said] that such emotional continuity is provided by good design, but no one can define good design. . . . . The efforts of the past we may admire and measure and classify, but we must speak of our own emotions in our own way, because the powers of the human mind draw strength from its own efforts and wilt from imitation. The creative powers of man are truly a divine gift. It is this creative desire to search for truth, even in a small measure and in his own inadequate way, which stimulates the architect to find new solutions. If he understands the importance of religious symbols as a means to historical continuity, such understanding will guide him and provide the discipline which must always be present in the work of any artistic importance.

From the above point of view, then, the problem facing the contemporary architect is not impossible of solution nor is it peculiar to his day. It is an old one, and each age has met it and solved it in its own way. With all the shortcomings of a materialistic world surrounding us, we too must face and solve it in the full realization that the main function of a church building is to provide emotional fulfillment.

Today's need for economy makes us avoid pompously designed monuments, and in so doing we have found that much significance can be imparted to simple materials such as wood or brick, and much warmth and feeling may be achieved by the judicious use of such intangibles as space, light, texture, and color. Paintings, sculpture, stained glass, and other decorative arts, if creative and not merely imitative, add immeasurably to the proper solution of the problem which, as I noted before, is to create an environment in which the average man may find spiritual shelter; a place where he may draw strength for his daily labors, and courage in his battle and temptations, a place where he may join others in worship and meditation.

I do not agree with the premise that if the liturgy is understood and appreciated, all other questions are readily solved, because the examples of many churches built in recent decades, while fulfilling all liturgical requirements, have failed to a great extent to create the emotional impact so necessary in the House of God.

I will admit that the task is more easily stated than carried out, due to the simple reason that there are few really creative minds. So I must end with what may seem an apology; the danger to contemporary religious architecture does not come so much from our right to express ourselves in a modern idiom, but from the fact that so few designers have the gift, the integrity, and the discipline to make such an idiom of convincing significance.

CHURCH ARCHITECTURE

Pietro Belluschi, Church Property Administration, *November-December 1949;* Architectural Forum, *December, 1949.*

The thoughtful architect will appraise the spirit which moved other ages, so that he may himself recapture such spirit, not by imitating, but by truly understanding it, which means seeing the thousand ties which bind architecture to its own age. These ties of course include among other things, people, labor, costs, materials, etc.; to follow that road is indeed to be in the tradition.

Unfortunately, our age is not a happy one; our society believes mainly in the importance of scientific progress, pays only lip service to the old images of God, and finds it difficult to formulate more convincing ones. It believes only in the material progress of men, which is brought about at the price of general unhappiness and wars. Our heaven is now on earth; it takes the shape of social security, the thirty-hour week, and restless and uncreative leisure —a heaven of course that gives neither serenity nor spiritual nourishment.

Then what is to be done about church architecture if we cannot find refuge in a sterile copying of the past? The answer does not rest in the mind of the few of us, but in the very fabric of our society. . . . . Many individuals have come to understand the predicament of our modern society, many architects have been baffled and impotent in their struggle to become again simple and believing; of these, only a handful have succeeded in recreating the atmosphere in which the religious man, the man who still deeply and humbly believes in God may worship Him in appropriate surroundings.

In spite of the advantages and satisfaction accruing to the modern religious congregation by its advocacy of social advances, it must be admitted that to many persons the results of this course is the least rewarding in spiritual satisfaction.

To them, God is still an intimate necessity, not satisfied by the knowledge that social advances have been made; to them, and to the many persons who are lonesome and bereaved, to the unhappy people whose only source of courage in their daily tribulations is their opportunity of prayer and emotional release;

to the sick and the dying, and the fearful to whom the last source
of strength is the image of a personal loving God; to all of those,
our modern religious establishments have to a large extent failed.
The interest which a community shows in its church is dependent
on the ability of the few to coax and corral funds from its members on the basis of social pressure; and in this case, the church
becomes the poor man's club, and the parish hall kitchen becomes
more important than the altar.

Then, the number of conflicting denominations has tremen-
dously weakened the church as a divinely mandated institution,
and the time and wealth which a community can devote to con-
struct its church could not possibly be compared with the wealth
devoted to the same purpose in the middle ages where life revolved
around the religious institutions. However, if we cannot erect
great monuments, we may endeavor to create small temples, in
a more human scale, designed in a sensitive and creative manner
so as to produce the kind of atmosphere most conducive to worship. This then is our answer.

Ray Atkeson

THE RESPONSIBILITY OF THE ARCHITECT IN THE WORLD OF TODAY

Pietro Belluschi, Lecture, University of Illinois, October 29, 1951.

In a civilization so oppressed by cosmic events, it is easy to become confused by the many voices which claim priority on our allegiance and our services. What do we expect of an architect? Is he to be just a good engineer and a master plumber, or a superior and perhaps arrogant artist—should he be merely a practical and reliable man, or above all, a man of vision.

Today the power of money, coupled with the finality of the useful, a Thomist philosopher once observed, have so ruined the leisure of the soul, and so shaped human activity in an inhuman way, that even "vision" has assumed the questionable connotation of daydream, and the man possessing it, is called *visionary* and is excluded so to speak from the table of the elected. Yet the architect must be one of the elected; to be effective, he must understand and be understood, and above all, be successful; so his education, which lasts a lifetime and has no end, must be designed to make him competent in many fields.

He must be a "businessman," which means having a knowledge of organization methods, of psychology and economy. He must be a "planner," which means knowledge of land economy, sociology, surveying, landscaping, etc. He must be an "engineer," which means knowing structure, as well as heating, ventilating, and electrical systems of distribution. He must be an "educated man," which means knowing and appreciating past and present cultures, and having more than a superficial interest in the arts of painting, sculpture, the drama, music, etc.

An architect must be all of these things and also be a sensitive, sensible man with sufficient honesty and integrity to inspire the full confidence of his clients. But I say that unless he is above all a man of vision, and an expert in the field of visual and spatial relationship; unless he is able to give form and order to space; in brief, unless he is a creative and understanding artist, he will not fulfill the peculiar role society expects of him.

He will be a leader only on that condition, because men of other talents can surpass him in all disciplines I have mentioned but that one.

His field of action as a creative man is the environment, large or small, be it a city or a home.

Failure to stand by that definition in the past has resulted in the general decline of our environment. We find even now that city planning is done by politicians and expediency experts, or at best, by sociologists and economists, who claim to know how to make a city work, but do not care whether it is emotionally fit to live in.

I claim that their city in the end will not work because the needed creative breath of life is not there, and people sense that.

If we agree then that the architect's role in our society is to bring orderly beauty where only useful disorder may exist, we may explore perhaps, and try to define, the standards by which he may be able to proceed in his task.

We may well admit that lately we have all become not only disbelieving and hesitant, but also a little nostalgic for the sensuous beauty of the very world which we had so eagerly discarded only a short while ago. It is possible for instance to detect different systems of approach in the works of the best architects of our time. We may observe how each of the great leaders has given different emphasis and unique interpretations to certain contemporary ideals. Each one has shown his own way of searching for beauty, and by his work has given us a glimpse of the eternal truth. The "romanticism" of Frank Lloyd Wright has shown us by its lyrical qualities to what heights of poetic power, emotion and imagination can lead architecture. Yet, while we need very much its stirring and quickening influence to inspire us in our daily work, we sense that it does not quite belong to our time nor to our society. His work has been the brilliant performance of a great artist, scornful of his age and intolerant of the people's problems. His imitators have also shown how forms conceived in the same spirit, but without the guidance of genius, can degenerate into undisciplined, shallow and arbitrary statements lacking poise or even plastic quality.

The "formalism" of Le Corbusier has given us back some of the order and serenity of the classical architectures of the past. Its concern for subtle proportions and plastic relationships has given us great delight and a feeling of scale, but the new formalism has shown tendencies of turning into academic mannerism, much in the same manner as the old formalism which lost its vitality when it crystallized into definite styles.

This trend has been prolific of many clichés in our architectural dictionary and it has so fascinated some designers that they found themselves using its forms even before the function of the building was established. We still remember the efforts of the eclectics to fit banks or office buildings into the skins of Greek temples, and the memory is not a pleasant one. Architectural forms which are not born of logic, study and deep understanding of the peculiar problem at hand, but come out of preconceived aesthetic theories will always be in danger of becoming artificial, tricky, or just fashionable.

Finally, we have "functionalism" as a clear influence on the thinking of many modern architects. At its best, such as in the works of Mies van der Rohe, it has given artistic expression to the products of the machine methods, and elevated them to the dignity of lasting architecture. Mies van der Rohe holds that technology is the great historical movement representing our epoch, shaping our world, and giving it unity and significance, just as religion did to the world of the middle ages. He goes as far as to say that architecture has nothing to do with the invention of forms, nor is it the playground of children, young or old.

While this may be a defensible philosophical theory, it has shown that in unskilled hands it can produce cold and inhuman interpretations, and we cannot help but feel that it will lead to intellectual puritanism and then to sterility.

It seems to me that technological methods are in a flux of continuous change and that truth is not always the expression of function because man himself is an exceedingly complicated machine for which no one has been able yet to lay down immutable laws or symbols. Yet, there is no question that "functionalism," by making us understand the aesthetic potentialities of our machine age and by showing us what can be accomplished by a severe discipline of approach, has fully justified itself and given us much from which to build.

It is interesting to observe that the major exponents of these trends have one thing in common, they are artists in the most exalted sense of the word and they are interested in the wider aspects of man's environment. They are men of vision, whose creative minds see spiritual relationships in the world of reality which escape less creative men. Their discoveries can be our beacons, yet each of us must chart his own course, must go his own way in the light of his own reason, and of his own ability to understand, because the living world has infinite variations and limitations which call for infinite variety of solutions and a continuously fresh approach. Each one of us needs the power of poetic expressions which he may use without compromising practical dictates; he must strive for formal order and proportion without being dominated by its formulas; he must understand the pervading power of technology without forgetting the primacy of the human soul and its thirst for sensuous variety; he must create with humility and discipline from what lives around him, look with awareness upon the "great American scene" and with sympathy and even love upon its people, believing that it falls on each one of us to work for their advancement and for the enjoyment and fulfillment of their daily lives. We have seen that

Ray Atkeson

24

architecture, in keeping with the social advances of our times, has changed its role mainly from one of service to the princes and the wealthy, to one of service to the common man, in endeavoring to improve his environment. That, in my opinion, is the greatest and most significant movement to which the modern architect is being called upon to take a part.

In that light and in that role there are being opened infinite sources of new and exciting adventures in modernism; and beauty, a virtue so difficult to define, yet so feminine in its willingness to give itself to those who care to search for it, will reveal itself in new and captivating, yet logical forms.

The new concepts of town organization, the new community facilities such as recreation, shopping, and health centers, parking, the green-belts and children's playgrounds; the new discoveries in the use of concrete and steel and wood and the use of the machine for mass production, which—while imposing limitations in design—makes possible the development and use of new materials and the great economies brought about by standardization; all these are the very essence of our search for modern architecture and it must be our purpose to give them aesthetic significance. While we may be discouraged at times by the slowness by which some of these changes come to pass, or by our inability to do our best, and while we may also feel nostalgic in the loss of our romantic past, we may draw courage from the wealth of opportunities opening before us—opportunities to use in a creative way our technical advances, to explore the needs and demands of a changing society, to help man to come to happy terms with his environment. If we are correct in such beliefs, we may then say that the achievements of our contemporaries must be judged by such broad standards, rather than by its externals or by the fashionable labels of modernism which up to now have been our measuring sticks. A house or a building is modern, not because it has a flat roof or a butterfly roof, continuous vertical spandrels, or horizontal spandrels, lally columns, or plastic bubbles, but rather because it has recognized the meaning of space in relation to its purpose, and to its setting; because, I repeat again, it has solved in a free, and creative way all the many social, economic, regional, emotional and practical limitations peculiar to the problem at hand. I want to make clear at this point that I do not minimize the importance of using fresh and original details in carrying through architectural works. They are the touchstone of competence, but just as literary works should be judged by their content and the depth of ideas expressed, rather than by the choice of words, so must the value of architecture be gauged by its deeper meaning which can then be expressed with "fantasy and imagination." On such ideas then, can our definition of *modern architecture* be based, superseding the old discredited definition which took the worst artificial and transitory features of architecture as its standard of judgment.

Excellent examples of modern architecture, understood in that manner, unfortunately are not many. To see the magnitude of the task still ahead of us we have only to cast our eyes about us. It would be easy to become discouraged by the outward manifestations of a society which places little premium on culture; yet, we may also sense all around us a great vitality, a desire for expression, a stubborn search for ideals and a thirst for beauty and sensuous satisfaction. It is up to the architects to give them form and fulfillment and to prove that the great powers of production of which our nation is so proud can be used to satisfy their spiritual as well as their physical needs—a balance which is so much needed for a happier society.

I hope that our democratic system for which at times we seem to pay such an extravagant price in terms of its inability to get things done, can rise to the challenge, without doing damage to our freedoms. Present dangers created by our external enemies will eventually disappear. It will be then that our ingenuity will be tested to see whether we can use production for the betterment of mankind, or only for its destruction. In that task the architect, together with the social scientist, the planner, the economist, and the engineer, will have a part, and to that end the education of our young architects should be directed.

Our immediate task, it seems to me, is to show our concern for the emotional needs of our clients and to show them that we are not reluctant nor unable to impart richness to the background of their lives, or to provide the kind of emotional fullness which played such an important role in the great periods of the past.

OUR READINESS FOR
BETTER ARCHITECTURE

Pietro Belluschi, Sunset Magazine, April, 1943.

The assumption that technical and scientific advances coupled with our ability to produce materials in great and varied forms will alone give us assurance of a fuller and more understanding life, or that by more machines the progress of civilization will be automatically stimulated, seems to me at least open to question. To believe that prefabrication and the use of light metals and even an awareness in community planning will, in themselves, give us beautiful architecture and happier surroundings, is to ignore the most important, yet probably the most neglected, factor which goes to make a civilized community: the individual himself, or what, for lack of a better name, we may call the "mass individual."

I deeply believe that all our technical improvements and all our efficiency and all our machines will bring us no nearer to a civilized society as a whole, unless a parallel effort is made to really educate the mass individual and to improve the whole system of human values and relationships. Recurrent wars testify to that belief. It would be easy to put the blame for our lack of spiritual awareness on many causes—we could even blame our schools for neglecting to turn our young students into sensitive people. It may be that we are trying to cram too much knowledge into their brains and not enough wisdom. It may be quite possible that the methods of imparting wisdom are lost to our age. But to hope for a civilization which will know how to make sensible use of its growing leisure without preparing for it seems to me forlorn.

In my opinion the house for which we should strain our energies, and into which we should place the seed of our hopes for the future, is a house which to the average individual will reflect a desire to live fully and with understanding, a house which will invite and inspire its owner to live in awareness and communion with his surroundings, that will make him a wiser human being and a better neighbor. . . . .

HOUSES

Pietro Belluschi, Lecture, Portland Art Museum, 1941.

The house must be a shelter, not only from the elements but also from a demanding and hard world. So we may say that the moral and intellectual capabilities of the individual are essential requisites to architectural achievements.

The real function of an architect is one of creative coordination. This task of coordinating has by degrees become more complex as the elements to be coordinated have become more numerous and scientifically more advanced.

Now, coordination presupposes a preliminary work of analysis and a final one of imaginative synthesis. What are the elements to be analyzed?

First of all, there is the architect himself; second, the person for whom the house is to be built; and third, the relationship between client and architect. If there is no basis for understanding, the whole enterprise is an unhappy one.

Fourth. The surrounding country—the form, color, and general aspects of the landscape on which the house is to be placed. A house may suit the mood of the countryside, the color of the soil, the shape of the trees, and the texture of the grasses, or it may dominate them.

Fifth. The orientation—not only the location of the house, but that of the windows and their shapes will be affected by this consideration.

Sixth. The climate will affect the materials which give the house its appearance, the roof which more than anything else affects its form.

Seventh. Surrounding buildings and the existence of a strong stylistic tradition. Their presence cannot be entirely disregarded. The new house as far as it is practical should be made to harmonize with its neighboring buildings, however without abdication of ideals.

Eighth. The methods of construction, which vary from region to region, and which depend on the habits of the local craftsmen, on the practical setup of the building machinery and on the custom of the people. If machine fabrication would really effect great economies, then it should be considered.

Ninth. The materials at hand.

Tenth. The financial restrictions of the client.

Eleventh. The physical requirements, based on the size of the client's family and on his standard of living and on his special hobbies.

Twelfth. All the mechanical equipment which has been created by this machine age and which is to be incorporated in the fabric of the house.

If we relate these different elements creatively in one unified whole, then we have modern architecture.

This concept of modern, therefore, will not lead us to expect it to be just another style. It cannot be labeled international style, although certain characteristics are universal; not modernistic. It should not even be called modern, because it goes back to fundamentals. It goes back to nature, if the owner's life is one of response to it. Therefore, we may deduct that a region with similar natural and human attributes may have an architecture harmonious to them. The people are neighbors, their interests are alike, they respond the same way to life, they have the same materials at hand, they have similar landscape, the same climate. So "regionalism" really has a meaning, which internationalism does not quite have.

PRINCIPLES AND TECHNIQUES
OF MODERN ARCHITECTURE

Pietro Belluschi, Journal of The A.I.A., September, 1951.

An unstable society such as ours cannot produce a Golden Age—all that we can hope is that this period be one of transition towards something better than we now have. True, we cannot discount the possibility that with our innocence we may have lost our ability to believe in ourselves, and thereby our chance to give unity to our own world; yet I, for one, believe that nothing can ever fully destroy our quest for truth in whatever form it is given us to see it—nor our desire to advance toward that great mirage which we call "beauty."

In fact, if we take the long look, we may see the human race again endeavoring against tremendous odds to find ways and means of harmonizing itself, as it has always really done in the past, with all the forces which surround it. This almost instinctive process of coming to terms with its environment, this desire to grow and become civilized by accepting and absorbing the many and complex challenges of its age, is what gives promise that significant and enduring architecture will yet emerge in our time. By the forms which are derived from our own needs and devices we shall be known to posterity, and it is by such standards that we should also try to judge contemporary architecture, and not merely by the externals, the fashionable clichés and the package embellishments which have become the labels of modernism. . . . . .

Man is a very complex animal, full of strange emotion, and illogical desires. He is swayed by prejudice, love, and hatred. He is a being pervaded by both idealistic and materialistic motives. He can be both gregarious and misanthropic, altruistic and selfish. No one can possibly know enough about man to draw immutable laws, sound conclusions or certainties; I think that planners and architects, just like politicians, must have their ears to the ground and listen, and understand and sympathize; they must interpret and lead, but not impose; their creativeness must spring from human understanding and even love. A painter, a sculptor, a composer may be haughty, detached, and even arrogant, but not an architect, because he has a social task to perform. Fortunately, we have increasing evidence that the modern architect is becoming more concerned with "man," and that "man" is again the criterion and the main motive on any level of planning.

The theory of planned neighborhoods, considered both as a standard of expansion and of rehabilitation, is based on this belief in the liberating influence on people of an environment which they can apprehend, welcome, and in which they can fully participate. . . . .

Beauty is yet our greatest motivating ideal, and the search for it our greatest source of strength. A plane in flight, the great suspension bridges, a high dam, a network of throughways cutting the landscape, the shopping centers, the green-belt cities—certainly these are new aspects of beauty more significant, more convincing to us than the old styles could ever be, because they belong to us, they are the symbols of our achievements. They also show promise that as we mature we may turn to other aspects of beauty equally fresh and equally ours. . . . . By that token we architects, of the common working variety, who must be front-line men, facing frustration and compromise; we, who must understand, absorb and give visual form to so many of the forces which make our world move, must not be ashamed to listen nor to understand what lives around us, ever mindful that each one of us can give more in a creative way by being part of the great mass of people, sharing their loves and enthusiasms, guiding them in the realization of their obscure ideals—not disdainful, temperamental stars—but men of vision among men. . . . .

If we bear in mind that our urban population has grown in the last five decades from one-third to two-thirds of our total population, the greatest concentration occurring in the metropolitan areas, we may understand the importance of our planning task. In 1950 these areas had twenty-seven percent of the nation's population and thirty-five percent of the effective buying power, yet the physical development problems due to this growth have been largely neglected or barely touched either by research or by action. We are beginning to realize with great impact that our cities are fast becoming obsolete, and that their potential rehabilitation is a process so long and so costly that there is literally no end in sight to the job which must eventually be done in order to make them again suitable to a better living. That is why I believe our job now to be so important.

If that is so, what then must be the task of our schools in educating new architects? Evidently they must be able to impart knowledge, but also wisdom. They must give the students awareness, but also doubt; they must give them discipline, but also freedom; the students must be taught to learn and remember and do, but also to think. We must, above all, encourage and develop their creative powers, because only by them will they be able to become leaders. Education must be the showing of the way to the never-ending search for cause and effect, for ideas, for sensual and visual satisfaction, for beauty as a living experience. . . . .

On one thing we may all agree: namely, that the architectural revolution of the first half of our century is over; our task now is reconstruction. We are finding that to rebuild is infinitely more difficult than to tear down.

SHOPPING CENTERS

From Hamlin, "Forms and Functions of Twentieth Century Architecture," published by Columbia University Press.

The market place as an institution for the exchange of goods goes back to the earliest days of history and has remained more or less unchanged through the centuries. It was one of the first manifestations of community life, curiously similar in the most widely separated parts of the globe. It may be said that a majority of the cities in Europe and Asia, as well as in the new continents, owe their existence to the early establishment of centers of trade. Such cities as Palmyra in Asia Minor, where camel caravans trading between the East and West could meet and exchange goods, offer a perfect example. The ample watered courtyard of the market place, surrounded by porticos affording the merchants shelter from rain and sun, was the prototype of what became later the forum, which in Rome was brought to a highly organized form. In the Roman Forum we find specialty shops as well as bazaars, arcades in addition to game areas and meeting halls; here people congregated for their various expressions of community living.

In the East, and in numerous European cities even now, the trading of goods of many types is still carried on in the bazaar or in the piazzas, either under the open skies or protected from sun or rain by porticos. In Spain, France, and Italy the cathedral square of the smaller towns is the place where, daily or weekly, merchants still display their goods in the shadow of the old church; here for centuries has been the center of community life, much as the forum was the community center in Roman times. In many of the larger cities some of the narrowest streets near the market place are the busiest trade centers, although vehicular traffic is prohibited. Arcades for pedestrians only, such as the glass-covered *gallerie* in Milan and Naples, have long been successful centers of specialized shopping.

Although this pattern of the market as a meeting place where goods are exchanged is still basically the same, the rise of large towns, the advent of mass factory production, and the need of obtaining supplies from wide national or international areas have tended to create complex problems of distribution with inevitable waste and paradoxically with increased efficiency. For instance the growth of rapid transportation created the necessity of bulk handling and led to the large terminal warehouse. This in turn necessitated the development of refrigeration. Efficient and economical marketing became a basic need out of which emerged the chain store.

The rapid urbanization of the last decades and, above all, the advent of the automobile have created extreme traffic congestion around downtown public markets. Decentralization of many types of retail shops has occurred in practically all the larger cities in the United States. The automobile has been responsible for creating conditions which, without guidance, have made merchandising a difficult task but have also made possible a solution. We see, therefore, the rise of the new concept, that of the shopping center, and indirectly that of the "neighborhood" as the cellular organization of the modern city. This concept has shown its maximum development near residential sections in the suburbs of large cities, and especially in the cities of the West, like Los Angeles, where the use of land still is not too highly concentrated. . . . .

It is impossible to lay down strict recommendations on the planning and design of the shopping center proper, because the variables are too great; these depend on the objectives of the management (quite different in each case), the area available, the climate, local customs, changing conditions, types of buying power, building codes, transportation facilities, labor conditions (as well as union conditions), the feasibility of self-service operations, highway systems, and many other factors, all of which affect the solution of a given problem. It may be said in general, however, that in designing a shopping center, the architect must consider the practical needs first and he must weigh carefully the cost of the investment against the financial returns.

A center designed as a group can be controlled and made attractive. People take pride in a well-designed and well-organized center. Several wartime projects demonstrated the architectural possibilities inherent in the problem; they showed, too, that the manner in which a center is planned and its development is controlled is instrumental in retaining the maximum amount of spending power of the community within the neighborhood. People learn to think of the shopping center as the focus and symbol of their community life, ... [Therefore] In designing a center it is important to strive for a harmonious architectural character. Harmony may be obtained by interestingly relating each building to the whole and also by the intelligent use of color, which in a dynamic way may help to integrate many otherwise unrelated forms.

The architectural effect of the center should be obtained primarily through an imaginative solution of the problem. As in all convincing architectural solutions, no practical usefulness should be sacrificed for mere architectural effect; but a sensitive use of materials, fixtures, and colors can give great distinction to a project without adding to its cost. . . . .

PART III

FROM BELLUSCHI'S WORK

Erven Jourdan

PORTLAND ART MUSEUM—1932 and 1939
The Portland Art Museum is made up of a block of three buildings facing on a park in downtown Portland, Oregon. The old building is still in use but is screened by the two handsome wings designed by Belluschi. The Ayer wing, the first to be constructed, comprises the entrance and sculpture court, as well as galleries. The Hirsch wing provides two floors of additional galleries. The early sketches show a gradual weeding out of many classical embellishments until the unadorned building remains to speak for itself. Belluschi's personal efforts have aided the museum organization in becoming a living part of the community life.

Roger Sturtevant

K. E. Richardson

K. E. Richardson

OREGONIAN BUILDING—1948

The large broad masses of the building for the Oregonian Publishing Company bespeaks the solidity of the century-old newspaper, and the position it holds in the area. The pale marble, red granite and green sweeps of glass give it color. The entrance façade is unobtrusive, but the main interest is to be found in the huge press room windows on the lower side, which produce a fascinating mechanical display day or night. The plan is carefully arranged to expedite the assembling and delivery of the newspaper, and includes a radio station and Hostess House.

K. E. Richardson

EQUITABLE BUILDING,
PORTLAND, OREGON—1948

An ethereal tower of sea green glass and
aluminum, is this building of the Equitable
Savings and Loan Association. The skele-
ton outline is predominant and stresses
neither the horizontal nor the vertical. The
rigid window pattern allows for more interior flexibility than is apparent as result
of arrangement of utilities and modular
subdivision. Reverse cycle systems of air
conditioning introduced in this and the
Oregonian Building are an engineering advance. The old Oregonian Building, which
no longer stands, can be noted in the left
background of the picture opposite.

Ezra Stoller

41

Ezra Stoller

Erven Jourdan

J. P. FINLEY & SON MORTUARY—1937
The Finley Mortuary is a new structure built around the old. Its most notable feature is the lofty and dignified Morning-light Chapel. There is no hint of denomination, but a definite feeling of spiritual serenity is achieved.

Erven Jourdan

OREGON STATE HOSPITAL,
SALEM, OREGON—1948

A treatment hospital for the mentally ill, it avoids an institutional feeling, is warmly colored and cheerful. Its semi-rural setting adds to the feeling of openness.

Boychuck

ST. THOMAS MORE CHAPEL—1941

The rural beauty of this once small church
has been partially obscured by the addition
of parish house and large school. The
humility and intimacy is recaptured, however, in the beamed interior. The smooth
warm woods are neither rustic nor formal.
Clerestory lighting of the altar is result of
the placement of the spire over the chancel,
rather than the more conventional position
over the entrance.

Erven Jourdan

Roger Sturtevant

ZION LUTHERAN CHURCH, PORTLAND, OREGON—1951

Here is another church with a village character, but with many new and effective features. The spire is there for its symbolic value, and the deep sheltering porch contributes to the impression of a haven to be found within. The warm rich colors, the many contrasting textures of wood, brick, glass and copper, and the emphasis of the unusual lighting are rewarding to see. The freestanding laminated wood arches support the roof and allow for aisles between them and the side walls. The copper relief of the entrance doors was executed by Frederic Littman.

CENTRAL LUTHERAN CHURCH,
PORTLAND, OREGON—1951

This church is more formal than the previous two, but the warm use of wood and brick and colored glass is still apparent.

Roger Sturtevant

CENTRAL LUTHERAN CHURCH (cont.)

The counterpoint of vertical lines and horizontal lines, of round and rectangular shapes, is delicately balanced. The bell tower is open on two sides and gains a look of strength without bulk. The round brick apse with its indented pattern of crosses, relieves the narrow windows and wooden mullions of the nave wall. The apse is

Roger Sturtevant

wider and higher than the nave and thereby allows clerestory and side light into the interior of the chancel.

The effectiveness of the lighting technique is apparent in the interior view. A reredos of wood and fabric screens the organ chamber. Broad laminated wood arches span the nave.

FIRST PRESBYTERIAN CHURCH,
COTTAGE GROVE, OREGON—1951
A sensitive congregation was willing to join
with Belluschi in exploring religious expression in terms of architecture, unfettered
by tradition, but without forsaking it.
There is no precedent for the gently rising
roof line, but it succeeds in focusing attention upon the sanctuary. Feelings are all
important in religious architecture, and
calculated but subtle details are essential
to provoke them.

K. E. Richardson

A deep covered walkway around the courtyard is an effective baffle between the mun-
dane world and the quiet world of meditation. The inscribed stone at the entrance
gate and the other carefully placed rocks
inside the court are reminiscent of the
sand gardens of the Ryuanji Temple in
Japan.

Julius Shulman

COTTAGE GROVE CHURCH (cont.)
The plainness of the plaster and light wood
nave is dispelled by the large window of
colored glass. The panes are so placed by
color and size as to make a patterned interplay of violet and yellow lights.

K. E. Richardson

The bell supports were carved by the pastor, the Reverend D. Hugh Peniston. The
lower view shows the window which il-
luminates a small chapel, and an entrance
to Sunday school rooms, the minister's
study, and recreational facilities.

Julius Shulman

Julius Shulman

Esther Born

Dearborn-Mas

LADD & BUSH BANK,
SALEM, OREGON—1941

The addition to the Ladd & Bush bank in
Salem is as simple as the original building,
by its side, is ornate. The cast iron molds
for the façade came around the Horn. However, the Victorian fancywork has gained
in dignity by the contrast to the smooth
marble and glass of its modern neighbor.

THE WHERRIE TAILORING SHOP FRONT
—1941

This building shows Belluschi's facility in
changing materials and mood—from the
formality of marble to the warmth of wood.
A limited budget is always a challenge to
his ingenuity, and many of his best efforts
are based on the beauty of the materials
themselves, as in the school shown at the
right.

SACRED HEART SCHOOL—1941

Dearborn-Massar

57

K. E. Richardson

FIRST NATIONAL BANK, SALEM,
OREGON, BRANCH—1946
The sculpture of Frederic Littman, in stone,
wood, or as copper bas relief, complement
many of Belluschi's designs. Eight scenes
of Northwest agriculture, industry and com-
merce are depicted on the front of this
Salem bank. Bronze accents the entrance
here, while colorful granite points up the
entrance of the Boilermakers Union Building (1942) shown on page 60.

K. E. Richardson

Dearborn-Massar

Photoart

BOILERMAKERS UNION BUILDING—1942

NORTHWEST AIRLINES,
PORTLAND, OREGON—1945

Shop fronts are primarily to attract attention, to identify the business conducted inside, and to guide the customer to and through the entrance. Such devices as the aluminum wall studded with round headed nails suggest an airplane; the pylon sign of Waddle's Restaurant (1945) attracts attention; the setback for easy parking is an invitation for customers to stop at the Electrical Distributing Inc. (1946). The back of this building is an honest, unadorned warehouse with loading platform.

Photoart

60

Dearborn-Massar

Dearborn-Massar

K. E. Richardson

PORTLAND OFFICE OF PIETRO BELLUSCHI —1947

Belluschi's Portland office is away from the center of town, and is a renovated garage with three levels. The uppermost level is a row of apartments over the main drafting room. The library, offices and utilities make up the split levels.

EDRIS MORRISON STUDIO—1947

The entrance court affords privacy to the patrons of the Edris Morrison photographic studio (below and opposite). The glass display screen effectively brings the advertising near the street.

Dearborn-Massar

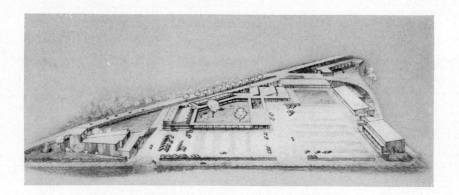

McLOUGHLIN HEIGHTS SHOPPING CENTER
—1942

Limited materials and wartime restrictions added to the problems of these shopping center projects. However, many valuable lessons were learned and a prototype established for future permanent centers.

BAGLEY & DOWNS SHOPPING
CENTER (opposite)—1942

Bagley Downs shopping center and temporary housing community for shipyard workers followed the very extensive McLoughlin Heights project. Both are located in Vancouver, Washington.

64

D. E. Edmundson

Dearborn-Massar

Dearborn-Massar

This Portland house is one of Belluschi's earliest (1938) and many people still consider it his best. It has a beautiful setting, an exhilarating view, and makes the most of both. The Japanese influence is present, and an over-all restraint is felt. It is built for a bachelor who had the architect design much of the furniture as well.

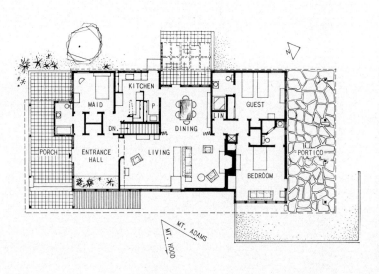

Boychuck

Boychuck

1938 PORTLAND HOUSE (cont.)

Boychuck

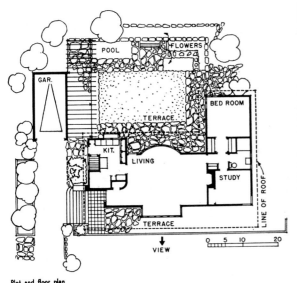

Plot and floor plan

Delano

Here is the first house Belluschi built for himself (1936). Its stark lines have been softened by the growth of the planting until it seems to repose peacefully on its site. A small home, it takes advantage of the level outdoor living area in the rear.

Boychuck

Delano

Dearborn-Massar

LINE OF FIRST FLOOR

STOR.

MAID
13x13

MAID
15'x13'

STOR.
22'x15'

UNFINISHED UPPER FLOOR

REGISTRES

B.R.
14'-6"
x12'

B.R.
14'-6"
x15'

BLANKET INSUL
L.R.
DROP IN CEILING
WOOD

D.R.

H.

VERANDA

K.

B.R.
12'x15'

P.   21'-8'

UTIL.

LINE OF ROOF OVERHANG

DOWN TO
HEAT. RM.

GAR.
21'-6"x20'

GROUND FLOOR

0 5 10 15 20 25 FT.

In a close view, this house (1941) blends
with the beach grass; from the long view,
the low roofline follows the horizon. The
sheltered look of the approach gives way
to a completely open face to the sea. Glass
and colorful stone piers compromise the
view wall. Furniture designed by Belluschi.

Dearborn-Massar

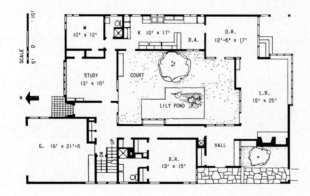

A sweeping view of Puget Sound from a Seattle, Washington, hilltop, is the special feature of this site (1941). For quieter beauty, an inner courtyard is contemplated from almost every room. Existing madrona trees became part of the design, and privacy and protection from winds from the Sound were problems effectively solved.

Dearborn-Massar

Dearborn-Massar

The outdoor areas of this Portland house
(1941) contribute greatly to its attractive-
ness and living enjoyment. Set in an apple
orchard, it boasts an extensive valley view,
but has its own closeup beauty in its rock
garden and its sheltered and open terraces.
The brick walk of the entrance porch continues into the interior hall.

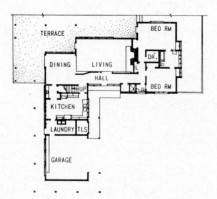

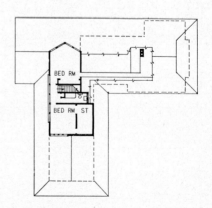

Dearborn-Massar

Dearborn-Massar

Informality and a view in two directions are notable features of this Portland house (1941). The naturalness of its setting, its many woods, and its structural nakedness give an air of rural simplicity, yet the house is only ten minutes from downtown.

K. E. Richardson

Delano

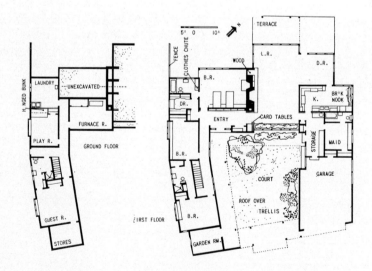

Dearborn-Massar

Facing Netarts Bay, an arm of the Pacific Ocean, is this yearround home (1942). The view side is the weather side, but a sheltered courtyard is formed by the bed-room and service wings. Woods are used profusely here in a decorative way as well as structurally.

Dearborn-Massar

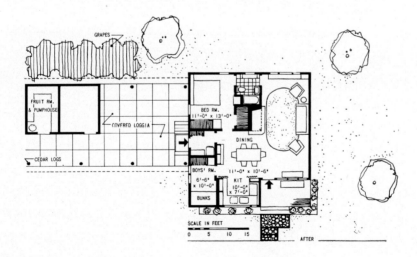

In remodeling an old farmhouse (1944)
for his own family use, Belluschi did the
usual things required to make it light and
open—removed partitions, added large
windows, and removed covered porches.
Then he added a huge concrete corner fireplace and had provincial furniture made by
craftsmen on the job.

Ezra Stoller

The compound of this ranch house (1948) encloses a rambling arrangement of office, spacious house, car shelter, and guest house —all under the same roof. The furniture was designed by the architect, the landscaping was planned by Thomas Church, and the copper hood fireplace relief was done by Frederic Littman. The workmen were skeptical of the flush fireplace wall, but eventually conceded to build it without indentation.

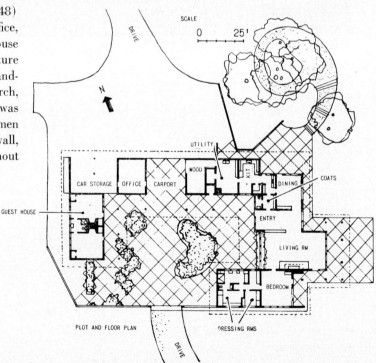

Ezra Stoller

A subtle blend with the arid terrain of
Eastern Oregon is carefully calculated here
(1948 house), and is a departure from the
forested and pitched-roof influence of the
Willamette Valley and coastal regions.
Here the black volcanic stone walls and
flat roof are at home beneath the rim rock;
the grey stained boards and battens have
the color of the sage.

Dearborn-Massar

The terrace faces the coolness of a fine fishing stream. The interior boasts three fireplaces for warmth in the cold and snowy winters. It is a sophisticated house, for all its rough materials. Rustic is a misnomer for any Belluschi house because it is invariably smooth and elegantly simple.

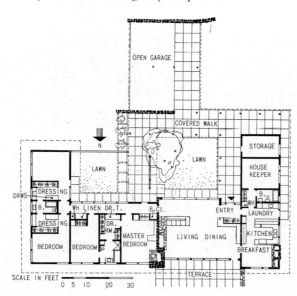

Dearborn-Massar

Dearborn-Massar

A smaller companion house to the one on
the preceding pages uses a rosy toned local
tufa rock with a sage green exterior stain
and chartreuse eaves. The colors are tied
into the decorating scheme and one short
wall of rough boards and battens continues
into the house.

Dearborn-Massar

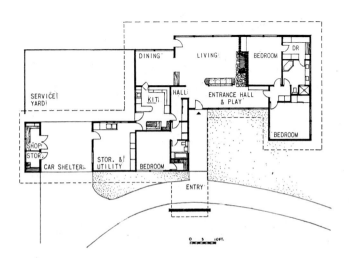

The hillside drops away sharply on the view
side of this Portland house (1949), and the
courtyard forms a baffle for privacy on the
entrance side—as well as a level and protected garden area. The simplicity of the
glazing detail is noteworthy.

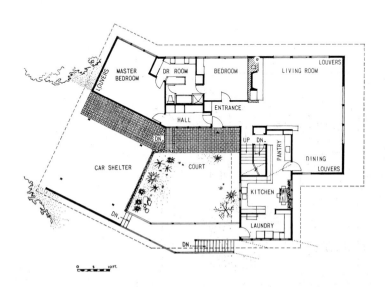

91

Roger Sturtevant

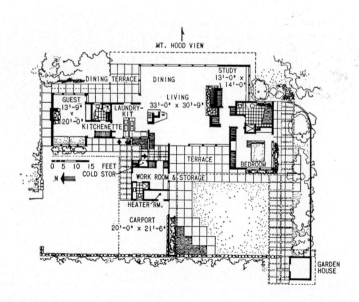

This seemingly secluded house (1949), in fact, is in a populous residential area, but most fortunately at the end of a dead-end street with no houses above it and a sharp drop on the view side. It is level enough for a spacious house and a large garden patio.

Roger Sturtevant

Roger Sturtevant

Perhaps this house will be considered Belluschi's closest brush with the so-called International Style, due largely to the particular taste of the client. The fetishes of an older mode of living were completely discarded and the freedom of this house is thoroughly enjoyed.

K. E. Richardson

This country home (1951, Orenco, Oregon)—a true farmhouse in looks and purpose—has the changing field approaching the house itself. Its two story windows look across farmland and woods to the mountains, and admit light to a large living room and balcony corridor.

BIBLIOGRAPHY

PAGE NOS.

AMERICAN HOME   *May 1944*   Beach House 20, 21
                         January 1945   Portland House 16, 17
                         February 1945   Beach House 42–45
                         February 1947   Farm House 30, 31

ARCHITECT AND ENGINEER   *October 1940*   Art Museum; Weiner's Store...... 33–35

                         March 1946   "Modern House Trends"......... 20, 21
                         June 1946   Shoe Store 19
                         April 1949   Equitable Bldg. 8–13

ARCHITECTURAL FORUM   *December 1937*   Finley's Mortuary 468–70
(The Magazine of Building) *October 1939*   Portland House 330
                         April 1942   Ladd and Bush Bank...........226, 227
                         May 1943   Office Bldg. 194X..............106–112
                         September 1943 Weiner's Store 71
                         May 1944   Shopping Center 95
                         February 1945   Portland House 101–105
                         June 1945   Civic Campaign Center 110
                         January 1946   Shopping Center 107–109
                         May 1946   Farm House 86
                                           Boilermaker's Bldg. 126, 127
                         April 1947   Equitable Bldg. 98–101
                         May 1948   Jewelry Store 131
                         July 1948   Medical Clinic 93
                         September 1948 Equitable Bldg. 98–106
                         December 1949   "An Architect's Challenge" (Churches) 72
                         January 1951   Zion Lutheran Church 142–146
                         May 1951   Portland House 163–169
                         November 1951   Portland House 226, 227
                         December 1951   Central Lutheran Church........163–167
                         May 1952   "Belluschi Appraises the Gropius
                                              Challenge" 113

ARCHITECTURAL RECORD   *February 1939*   Portland House 56–58
                         November 1939   Portland House 94–97
                         January 1940   Art Museum Wing 29–34
                         February 1941   Willamette Universary Library.... 85
                                           Northwest Airlines 118
                         October 1942   Shopping Center 66, 67
                         September 1945 Methodist Chapel 102
                         May 1946   Seaside Apartments 77–79
                         February 1948   Apts. Over Office 120, 121
                         October 1950   Oregon State Hospital 130–135
                         July 1951   Two Companion Houses 100–109

(California)
ARTS & ARCHITECTURE   *December 1940*   "An Eastern Critic Looks at Western
                                              Architecture" ....

                         January 1944   Four Houses ....
                         January 1945   Portland House ....
                         April 1949   Edris Morrison Studios ....

CHURCH PROPERTY   *May–June 1945*   St. Thomas More Chapel......... 18, 19
   ADMINISTRATION   *Nov.–Dec. 1949* Trend of Church Arch. 16, 17, 75

CIVIL ENGINEERING   *July 1950*   Federal Reserve Bank;
                                           Earthquake Resistant 38

CONSTRUCTION NEWS   October 21, 1950   Multnomah Bank 10, 11
  BULLETIN   March 17, 1951   Central Lutheran Church 8–10

ENGINEERING NEWS-RECORD   Feb. 3, 1949   Equitable Heat Pump 65–68

HOUSE BEAUTIFUL   December 1946   Barn Influence 160
                          August 1947   Seattle House 30
                          September 1947   Seattle House 106

INTERIORS   August 1943   Portland C. of C. 45
                          October 1944   Beauty Parlor 76
                          August 1945   Methodist Chapel 71
                          August 1946   Portland House 76
                          August 1949   Ranch House 81
                          December 1949   Equitable Bldg. 71

LITURGICAL ARTS   May 1949   St. Thomas Moore Chapel and School 92, 93
                          November 1950   Church Architecture ....

MORTUARY MANAGEMENT   September 1937   Finley's Mortuary 11–16

PACIFIC BUILDER AND
  ENGINEER   October 15, 1938   Finley's Mortuary 3, 4

PENCIL POINTS   July 1942   St. Thomas More 59–75
(Progressive Architecture)   Houses, Monograph
                          February 1944   Shopping Center 50, 51
                          August 1944   Shoe Store 57
                                              Beauty Parlor 46, 47
                          December 1944   Beach House 68–74
                                              Sacred Heart School 77–82
                          January 1945   Orchard Heights Theater 50, 51
                          October 1945   Northwest Airlines 71–73
                          May 1946   Beach House 76–80
                          July 1946   Shopfront 45
                          June 1947   Waddle's Drive-In 61–63
                          August 1947   Church of the People........... 60, 61
                          March 1948   Electrical Distributing Co........ 58–60
                          April 1948   Country house 73, 74
                          February 1949   Case study 39–54
                          June 1949   P/A Awards: Ranch house 61–66
                                                          Equitable bldg. 54
                          January 1951   Marion County Courthouse....... 46
                                              Cottage Grove Church........... 75
                          March 1952   Cottage Grove Church........... 63–68
                          August 1952   Bath-Dressing Room 119

SUNSET MAGAZINE   April 1943   "Readiness for Better Architecture" 10, 11
                                              Seattle house 12, 13
                          August 1945   Farmhouse 12
                          November 1947   Ranch house 50
                          February 1948   Portland house 48
                          January 1950   Portland house 38

Hamlin, Talbot: FORMS AND FUNCTIONS OF TWENTIETH CENTURY ARCHITECTURE, Columbia University Press, "Shopping Centers," by Pietro Belluschi.

Ford, James, and Ford, Katherine Morrow: THE MODERN HOUSE IN AMERICA, Architectural Book Publishing Company, page 123.